Handwritten Notes For Botany...

Plant Growth and Growth Regulators

V. Darani M.Sc., M.Phil., SET

Chapters Enclosed....

- Chapter 1: Plant Growth
- Chapter 3: Auxins
- Chapter 4: Gibberellins
- Chapter 5: Cytokinins
- Chapter 6: Ethylene
- Chapter 7: Abscisic acid
- Chapter 8: Brassinosteroids
- Chapter 9: Morphactins
- Chapter 10: Traumatic acid
- References

Chapter-1
Plant Growth

> Growth is defined as the permanent and irreversible increase in size, length and volume of plants. Growth is confined to root apex, shoot apex and vascular cambium in plants.

Undifferentiated masses of cells called meristems occupy this growing points. Growth of root and shoot apices results in increase in length of the plant whereas the growth of vascular cambium increases the girth of the plant.

The activity of each meristem influences the activity of other meristems near to it which results in growth correlations. For example, while the main apical shoot meristem is active it retards the activity of recently initiated lateral bud meristems. This condition is termed as apical dominance.

Growth is one of the most fundamental and conspicuous characteristics of living being. Growth is generally accompanied by metabolic processes (both anabolic and catabolic), that occur at the expense of energy.

2 | Plant Growth and Growth regulators

Growth is the outcome of cell division, enlargement of new cells and their differentiation into different tissues. The cell number is increased by repeated Mitotic divisions of the cells in the Meristematic region. The newly formed cells increase in volume by enlargement brought about by the synthesis of Protoplasmic constituents and cell wall materials. Finally, the enlarged cells undergo differentiation into various tissues. As a result of growth, the fresh weight of the plant increases day by day.

Plant growth - Indeterminate

Plant growth is unique because the plants retain capacity for unlimited growth throughout the life and is brought about by apical Meristems. The cells of such Meristems have the capacity to divide and self-perpetuate. The product, however soon loses the capacity to divide and such cells make up the plant body. This form of growth wherein new cells are being continuously added to the plant body by the activity of Meristem is called the open form of growth.

Apical Meristems are mainly responsible for the primary growth of the plants and principally

Plant Growth and Growth regulators

contribute to the elongation of the plants along their axis. The secondary growth of the plant is produced by the vascular cambium and cork cambium which results in increase in girth of the plant.

Fig: Diagrammatic Representation of locations of growth Points

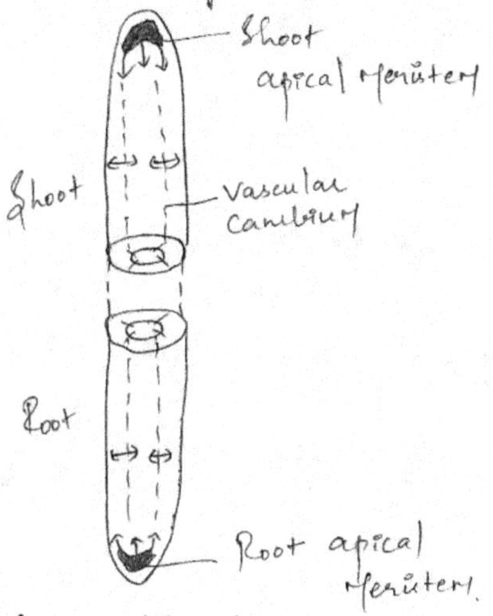

- Shoot apical Meristem
- Shoot
- Vascular Cambium
- Root
- Root apical Meristem

Growth - Measurable:

At cellular level, growth is a consequence of increase in the amount of Protoplasm. Since it is difficult to measure the increase in Protoplasm directly, one generally measures some quantity which is more or less proportional to it. Hence growth is measured by a variety of parameters like, increase

4 | Plant Growth and Growth regulators

in fresh weight, dry weight, length, area, volume and cell number.

Phases of growth:

The period of growth is generally divided into 3 phases namely, Meristematic, elongation and Maturation.

The constantly dividing cells at the root and shoot apex constitute the Meristematic phase of growth. The cells in this region are rich in protoplasm, possess large conspicuous nuclei. Their cell wall is primary in nature, thin and cellulosic with abundant Plasmodesmatal connections. The proximal to the Meristema zone represent the phase of elongation. Increased vacuolation, cell enlargement and new cell wall deposition are the characteristics of the cells in this phase. Further away from the apex is the portion of axes which is undergoing phase of Maturation. The cells of this zone attain their maximal size in terms of wall thickening and protoplasmic modifications.

Growth Rates:

The increased growth per unit time is called growth rate. Thus rate of growth can be expressed

Plant Growth and Growth regulators

Mathematically.

Fig: Arithmetic and Geometric growth

a) Arithmetic   b) Geometric

Fig:- Stages during embryo development showing geometric and arithmatic phases

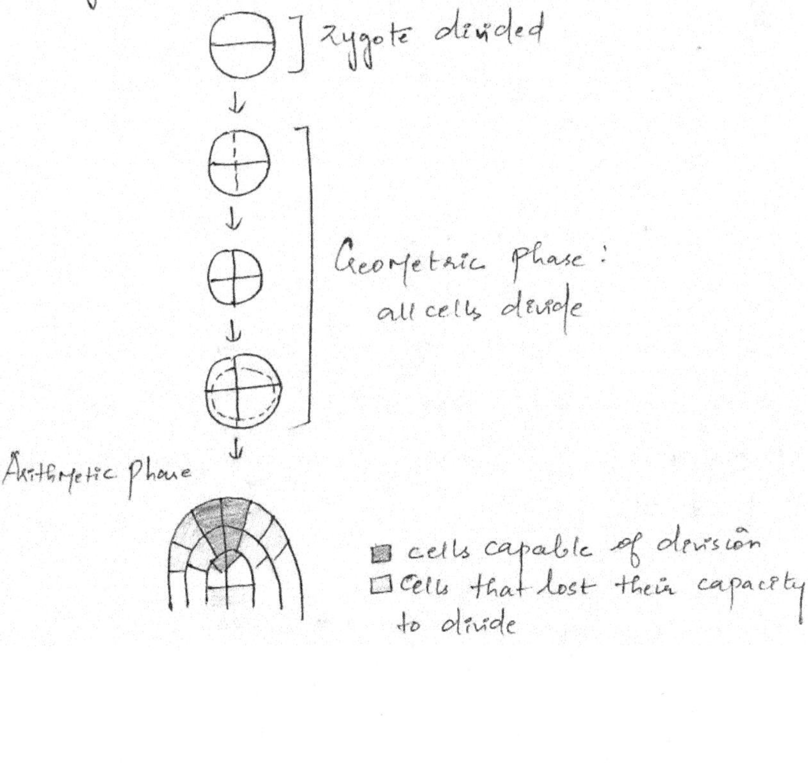

Zygote divided

Geometric phase: all cells divide

Arithmetic phase

■ cells capable of devision
□ Cells that lost their capacity to divide

The growth rate shows an increase that may be arithmetic or geometrical.

In arithmetic growth, following mitotic cell division, only one daughter cell continues to divide while the others differentiate and matures. The simplest expression of arithmetic growth is exemplified by a root elongating at a constant rate.

Fig: Constant linear growth

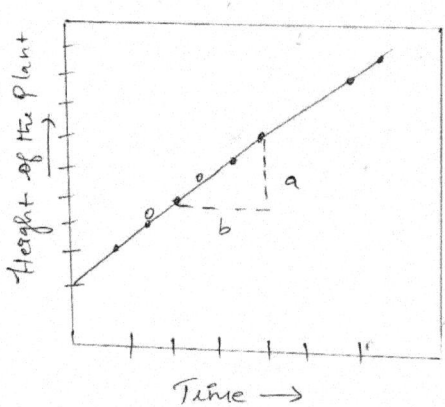

On plotting the length of the organ against time, a linear curve is obtained. Mathematically, it is expressed as,

$$L_t = L_0 + rt$$

L_t = Length at time 't'
L_0 = length at time 'zero'.
r = growth rate / elongation per unit time.

Plant Growth and Growth regulators

In geometrical growth, the initial growth is slow (lag phase), and it increases rapidly thereafter - an exponential rate (log or exponential phase). Here, both the progeny cells following mitotic cell division retain the ability to divide and continue to do so. With limited nutrient supply, the growth slows down leading to a stationary phase. On plotting the parameter of growth against time, a typical sigmoid or S-curve is obtained.

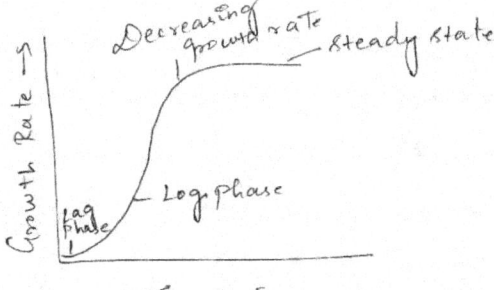

A sigmoid curve is the characteristic of living organism growing in natural environment. The exponential growth can be expressed as,

$$W_1 = W_0 e^{rt}$$

$W_1 \rightarrow$ final size (weight, height, number etc)
$W_0 \rightarrow$ initial size at the beginning of the period
$r \rightarrow$ growth rate
$t \rightarrow$ time of growth
$e \rightarrow$ base of natural logarithms.

Here, r is the relative growth rate and is also the measure of the ability of the plant to produce new plant material, referred to as efficiency index. Hence the final size of W_1 depends on the initial size, W_0.

In unicellular organisms like bacteria, growth is assessed by a count of number of cells per millilitre at increasing times after the cells are placed in a fresh nutrient medium and under environmental conditions (light, temperature etc.) suitable for optimal growth. There is an initial lag period during which cells activate their biochemical machinery for rapid growth by producing necessary enzymes. This is followed by a time period during which there is an exponential increase in cell number which is called as log period. This period of rapid growth does not continue indefinitely and due to depleted nutrient supply, accumulation of toxic products and other limiting factors ultimately leads to decreasing cell number until the population of cells reaches a steady state in which the number of cells remains constant (stationary) or even declines. If number of cells per millilitre is plotted against time (hours), again a sigmoid curve is obtained.

Plant Growth and Growth regulators

The ratio of the change in cell number (dn) over the time interval (dt) is called as absolute growth rate (AGR):

$$AGR = dn/dt$$

The AGR when divided by total number of cells present in the Medium, gives relative growth rate (RGR):

$$RGR = AGR/n$$

AGR and RGR are useful because they help to describe dynamics of cell growth in culture.

Conditions for growth:

Water, oxygen and nutrients are very essential for growth. The cells grow in size by enlargement which in turn requires water. Turgidity of the cells helps in extension growth. Thus plant growth and further development is intimately linked to water status of the plant. Water also provides Medium for the enzymatic activities needed for growth. Oxygen helps in releasing Metabolic energy needed for growth activities. Nutrients (Macro and Micro essential elements) are required by plants for the synthesis of Protoplasm and act as a source of energy.

Every plant has an optimum temperature best suited for its growth. Any deviation from this range could be negative to plant growth.

Differentiation, Dedifferentiation and Redifferentiation

The cells divided from the shoot, root apical Meristems and Vascular cambium differentiate and Mature to perform various functions. This act leading to maturation is called differentiation. During this, the cells undergo some major structural change both in their cell wall and protoplasm.

Plants show another peculiar and interesting phenomenon. The living differentiated cells that have lost their capacity to divide can again regain the capacity of division under certain conditions. This phenomenon is termed as dedifferentiation. While doing so, such cells are able to divide and produce new cells which once again lose their capacity to divide but Mature to perform specific functions (ie) redifferentiated.

Measurement of Growth:

Growth in plants can be measured in terms of either (1) an increase in length or girth as in stem or root (2) change in size (area) in case of leaves and flowers (3) Change in number of cells (4) Volume change in course of time in case of

Plant Growth and Growth regulators

fruits. The following methods are usually employed for measuring the growth of the plant:

➤ Direct Method:

This is the most simple rather crude method of measurement of growth in which the length of the growing part is measured just with the help of scales after intervals.

➤ Horizontal Microscope:

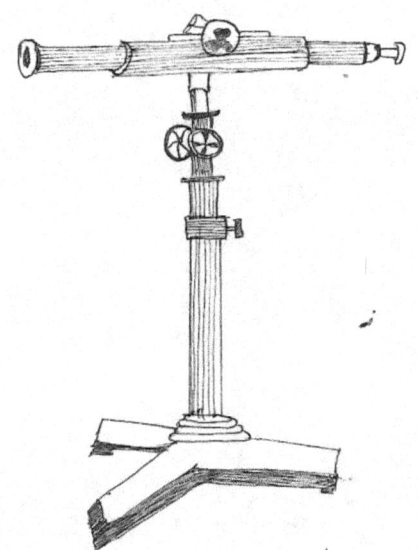

It is a simple compound microscope fitted horizontally on a stage that can slide over a graduated vertical stand. In this method, a point is marked near the stem or root tip and is focussed by horizontal Microscope. After some time, the same point is focussed either by raising (stem tip) or lowering (root tip) the horizontal Microscope. The difference of the initial and final readings on the graduated vertical stand measures the growth of stem or the root tip in length.

3. Arc Auxanometer

Arc Auxanometer is an instrument to measure the growth of the plant. It consists of a pulley connected to a pointer on an arc scale fixed on a vertical stand

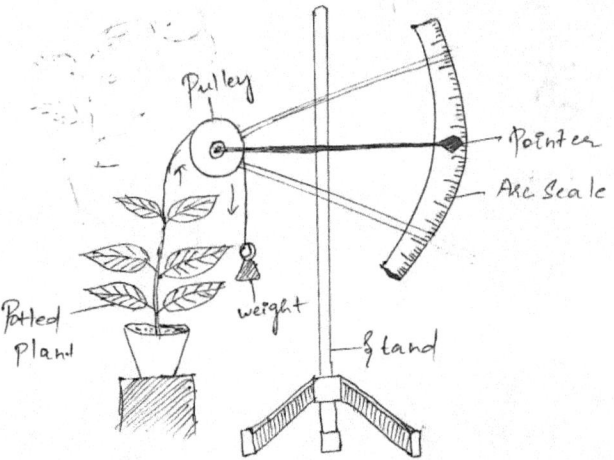

A thread can be passed over the pulley with one end tied to the growing point of the plant and the other end is tied with a light weight to keep the thread stretched. As the stem tip increases in length, the pulley moves and the pointer slides over the graduated arc. The reading is noted. Actual increase in length of the plant is then calculated by knowing the length of the Pointer and diameter of the Pulley.

➤ Pfeffer's Automatic Auxanometer

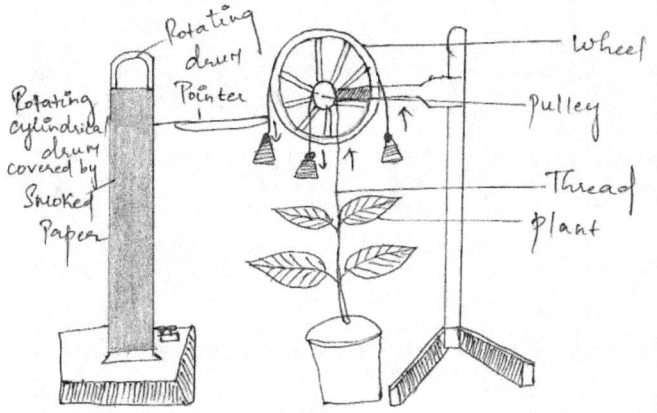

This consists of a vertical stand with two Pulleys. A thread is passed over one pulley and a small weight is tied at either ends of the thread to keep it stretched. A fine pointer is attached with one side of the thread. The Pointer is kept in touch with a rotating drum smoked on outer surface.

Another thread whose one end is tied to a small weight and the other end is tied to the tip of the plant, is passed over the other pulley.

As soon as the stem tip increases in length, the wheels of the Pulley move so that the pointer also moves downward and traces a special white marking on the smoked paper which is the indicator of growth of the stem tip. Actual increase in length of the stem can be calculated by knowing the radii of larger and smaller wheels and the rate of rotation of the drum.

➤ Bose's Cresograph:

This sensitive instrument is used to measure the growth of compound leaves. This instrument is similar to Pfeffler's auxanometer but the rotating drum is replaced by a glass plate that oscillates to and fro at regular interval.

Factors affecting growth

The factors that affect growth are as follows:

➤ Food supply:

The food supply is directly proportional to the rate of growth. With deficient food supply, the rate of growth slows down and ultimately stops.

➤ Water supply:

Supply of water is also directly proportional to the rate of growth since it directly affects the protoplasm activities.

➂ Temperature:

Optimum temperature is needed for good growth and it varies with the kind of plants.

➂ Light:

Light affect plant growth variously due to the difference in intensity, quality and periodicity.

Light intensity → High intensity increases the rate of water loss which in turn reduces growth. Low intensity reduces the overall growth by reducing photosynthesis.

Quality of light → Different wavelengths of light affects the plant growth in various ways. The red light favours the growth, the infrared and UV light retard the growth.

Duration of light → Flowering is dependent on the duration of day and night hours.

➂ Oxygen supply:

Oxygen increases growth. It helps for respiration, a process that converts potential energy into kinetic energy which is needed for various metabolic activities of the plant. Thus oxygen induces growth.

Chapter-2
Growth Hormones

> Organic substances which are produced in one part of the plant and transported to another part where, in minute amounts, they affect growth are called growth hormones.

As growth hormones regulate the growth related activities of plants, they are popularly known as growth regulators. The plant growth regulators are small, simple molecules of diverse chemical composition. Various physiological activities and growth in plants are regulated by the action and interaction of these chemical substances.

The plant growth hormones are variously called as plant growth regulators, plant growth substances or phytohormones. Plant growth regulators are broadly classified into two groups.

→ One group of PGRs are involved in growth promoting activities like cell division, cell enlargement etc. Examples — auxins, gibberellins, cytokinins.

→ Other group of PGRs are involved in plant responses to wound and stresses of biotic and abiotic origin. They are also involved in various growth inhibiting activities like dormancy and abscission.

Example: Ethylene, abscisic acid

Pincus and Thimann in 1948 defined a plant hormone as "organic substance produced naturally in the higher plants, controlling growth or other physiological functions at a site remote from its place of production and active in minute amounts."

Thimann in 1948 proposed the term Phytohormone for these hormones. They are of many groups like,

1) Auxin
2) Gibberellins
3) Cytokinins
4) Abscisic acid
5) Ethylene
6) Morphactin

Plant Growth and Growth regulators

Auxin

Auxins are growth hormones which in a small concentration bring about cell elongation in shoots.

The term auxin was introduced by Kogl and Haagen Smit in 1931. The discovery of auxin dates back to 19th century when Charles Darwin was studying tropisms in plants. He noted that the grass coleoptile when exposed to unilateral light bend towards light. He then covered the coleoptile tip with foil and cut it off from the unilateral light, the tip didn't bend. He concluded from this experiment that some 'stimulus' is transmitted from upper to lower part of the coleoptile which induced bending.

The explanation regarding this bending was given by Paal in 1919. He cut off the coleoptile tip and replaced it asymmetrically on the cut coleoptile stump and found that the coleoptile bent away from the side bearing tip even in dark. Thus he concluded that the tip secretes a substance that promote the growth of the part below it. The growth is symmetrical when the tip is intact and receiving light from all the sides. Asymmetrical growth results from the asymmetrical distribution of this growth substance.

Fig: Darwin's Discovery

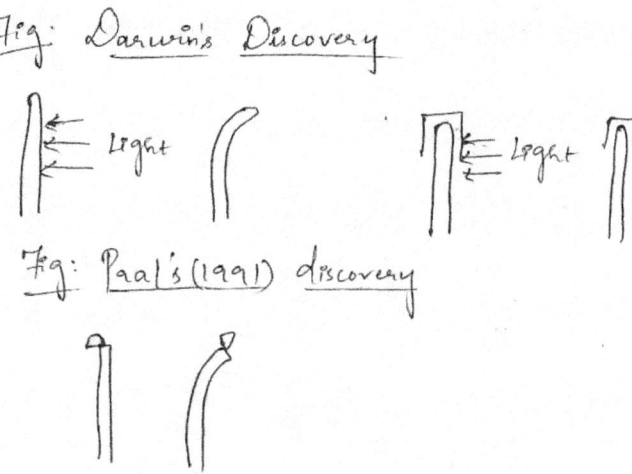

Fig: Paal's (1991) discovery

F.W. Went (1926, 1928) was successful in isolating this growth-substance from <u>Avena</u> coleoptile tips which still retained the growth promoting activity. He cut off the tips of the coleoptiles and placed them on small agar-blocks for certain period of time and then placed the agar blocks asymmetrically on cut coleoptile stumps. All the coleoptiles show typical curvature in the dark.

Kogl and Haagen Smit in 1931 isolated an active substance from human urine which was called as Auxin-A (Auxen triolic acid). Later, a similar substances was isolated from Malt and corn grain oil and was named Auxin-B (Auxenolonic acid).

Fig: Auxin A (Auxentriolic acid)

$$C_2H_5-CH(CH_3)-CH(CH_2-)-C(=)-CHOH\cdot CH_2\cdot CHOH\cdot CHOH\cdot COOH$$

$$C_2H_5-CH(CH_3)-CH(-)-CH=CH$$

Fig: Auxin B (Auxenolok acid)

$$C_2H_5-CH(CH_3)-CH(CH_2-)-C(=)-CHOH\cdot CH_2\cdot COCH_2\cdot COOH$$

$$C_2H_5-CH(CH_3)-CH(-)-CH-CH$$

Kogl et al. re-examined human urine in 1934. Thimann in 1935 examined Rhizopus culture and isolated a different substance which was named as heteroauxin. It appeared to be identical with an earlier known compound called Indole 3 Acetic acid (IAA). IAA has now been identified in a great variety of higher plants.

Some naturally occurring auxins are:
→ Indole-3-acetic acid (IAA)

- Indole-3-acetaldehyde
- Indole-3-Pyruvic acid (IPA)
- Indole-3-ethanol

Synthetic auxins are as follows:
- Indole butyric acid
- 2,4-Dichloro Phenoxy acetic acid (2,4-D)
- Naphthalene acetic acid (NAA)

Fig: Indole-3-acetic acid (IAA)

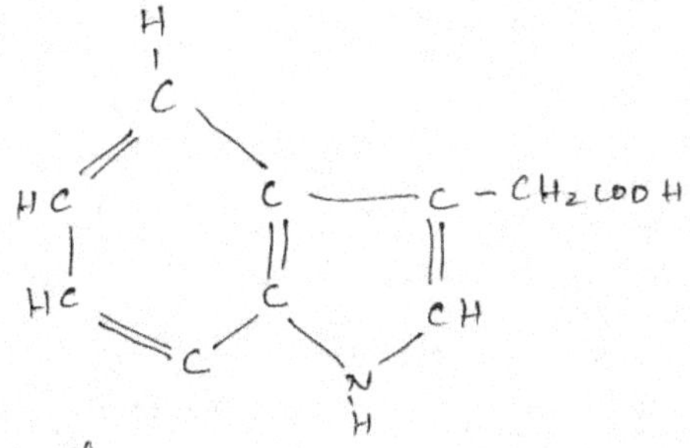

Indole-3-butyric acid (IBA) 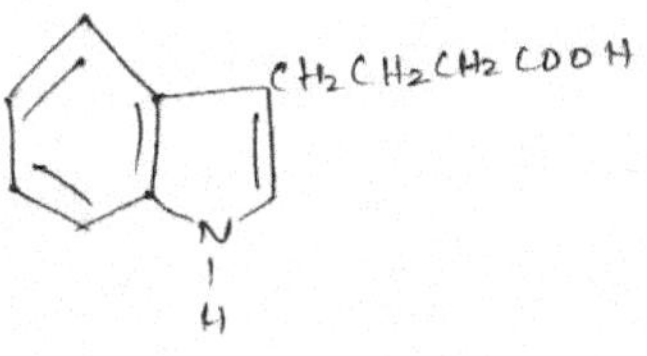   4-chloro-Indole-3-acetic acid

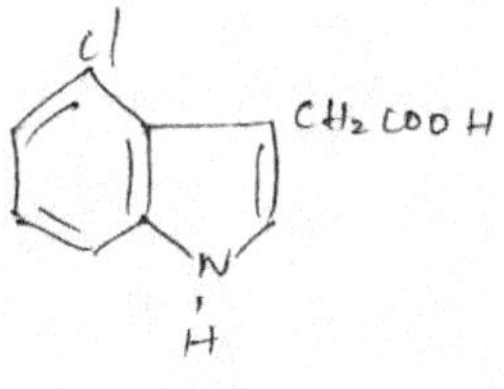

Chemical Nature of Auxins

Chemically Indole acetic acid (IAA) consists of an indole nucleus and acetic acid side chain. During metabolism, all the forms of auxins are converted to IAA.

Free auxins exist as free soluble auxins hence can be easily extracted from the plants by using organic solvents. They are biologically active auxins. The bound auxins remain inactive in cell and exist in conjugation with sugars, sugar alcohols or proteins.

The conversion of free auxin to bound auxin provides a way to maintain proper IAA concentration in plants.

Distribution of IAA in plants

Auxin is widely distributed throughout the plant but its concentration varies with parts of the plant. IAA is synthesized in the growing tips of plants from where it is transported to other parts of the plant. Hence the concentration of IAA would be more in shoot and root tips (in short Meristem)

<u>Dicot seedling</u> → In dicot plants, IAA is found to be highly concentrated in the growing regions of shoot, root, young leaves and developing axillary shoots. High concentration is found in

the growing tip of the shoot and decreases gradually towards the base.

Monocot seedling → In monocot seedling high concentration of auxin is found at the tip of coleoptile and decreases gradually towards the base.

In dicot seedling, the concentration of IAA gradually increases from the base of the stem to the tip of roots. In the case of monocot seedling, the concentration of IAA declines gradually from the base of the coleoptile to the tip of the root.

Fig: Relative concentration of auxin in different plant parts

10 = highest   1 = lowest

Monocot seedling

10
8
6
4
2
3
5
7

Dicot seedling

10 8
      6
2
3   4
9
2
5
7

Biosynthesis of Auxin (IAA) in plants

The synthesis of IAA occurs in shoot buds, young leaves, developing fruits and root tip. There are 4 Pathways for the synthesis of IAA from tryptophan. They are IPA pathway, TAM pathway, IAN pathway and IAM pathway.

Fig: Pathways of IAA Synthesis

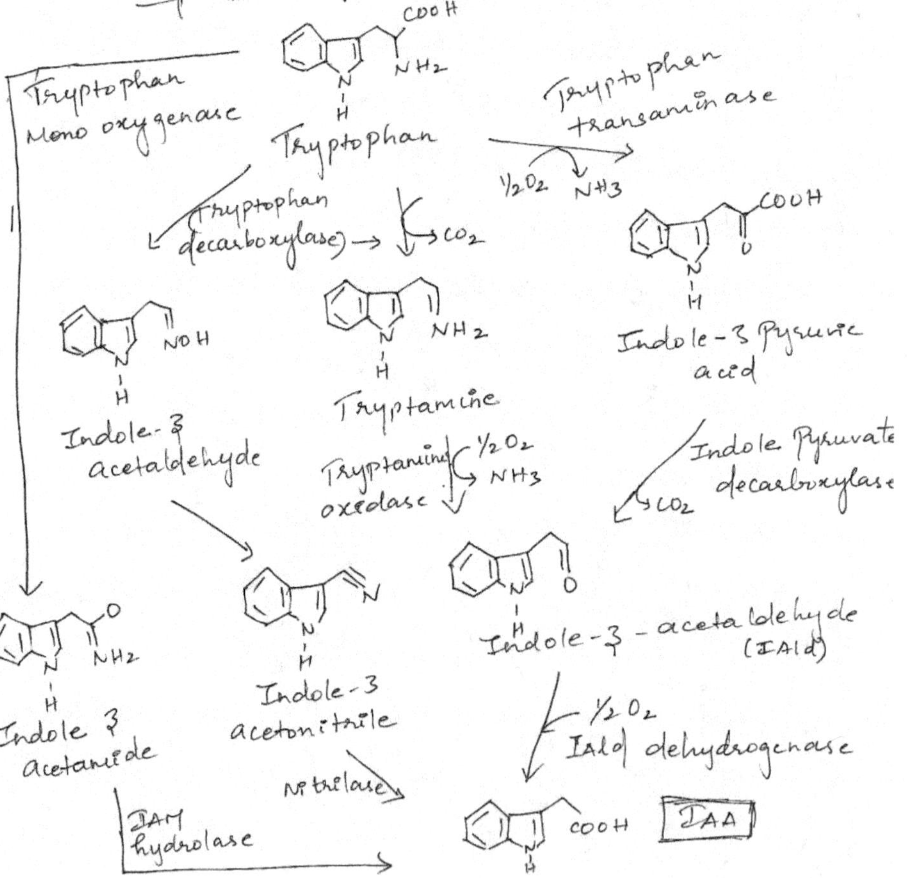

[Diagram showing the four pathways of IAA biosynthesis from tryptophan:

- Tryptophan (central starting compound, with COOH and NH₂ groups)

- **Pathway 1 (left):** Tryptophan → (Tryptophan Mono oxygenase) → Indole-3 acetaldoxime (NOH) → Indole-3 acetaldehyde → Indole-3 acetamide (NH₂)

- **Pathway 2:** Tryptophan → (Tryptophan decarboxylase, −CO₂) → Tryptamine (NH₂) → (Tryptamine oxidase, ½O₂, NH₃) → Indole-3 acetonitrile → (Nitrilase) → IAM hydrolase

- **Pathway 3 (right):** Tryptophan → (Tryptophan transaminase, ½O₂, NH₃) → Indole-3 Pyruvic acid (COOH) → (Indole Pyruvate decarboxylase, −CO₂) → Indole-3-acetaldehyde (IAld) → (½O₂, IAld dehydrogenase) → **IAA** (COOH)]

IPA Pathway

→ Tryptophan deaminated into Indole-3 Pyruvic acid catalyzed by enzyme tryptophan transaminase.

→ Indole-3 Pyruvic acid is decarboxylated to Indole 3 acetaldehyde catalyzed by enzyme IPA decarboxylase

→ Indole-3 acetaldehyde is oxidized to IAA. catalyzed by enzyme IAld dehydrogenase

TAM Pathway

→ Tryptophan is decarboxylated to tryptamine (TAM) by enzyme tryptophan decarboxylase

→ TAM is deaminated to Indole-3 acetaldehyde by enzyme tryptamine oxidase.

→ Indole 3 acetaldehyde is oxidized to IAA by enzyme IAld dehydrogenase

IAN Pathway

→ Tryptophan is deaminated by the enzyme tryptophan transaminase

→ This deaminated tryptophan is then decarboxylated to Indole 3-acetaldoxine by enzyme tryptophan decarboxylase

→ Indole-3-acetaldoxine is then converted to Indole-3-acetonitrile.

→ Enzyme nitrilase hydrolase removes nitrile group from Indole-3-acetonitrile to form IAA

<u>IAM Pathway</u>:
→ Tryptophan is converted to Indole-3-acetamide by enzyme Tryptophan monooxygenase.
→ Indole-3-acetamide is then converted to IAA by IAM hydrolase

⇒ <u>Transport of Auxins in plants</u>

Auxins as mentioned earlier are synthesized in the growing tips of root and shoot and transported to other regions of plants. Hence, it is a polar transport.

The transport of auxin in stems is basipetal polar transport since it takes place from the apex towards the base. Whereas the transport of auxin in root is acropetal polar transport since it occurs from tips of root to the base of the stem.

The most accepted mechanism explaining polar transport of auxin is Chemo-osmotic theory. According to this theory, auxins may be transported in any direction and any

Plant Growth and Growth regulators

Cell can uptake auxin from adjacent cells. The driving force for auxin transport is provided by proton-motive force across the membrane. Steps involved in chemo-osmotic theory are as follows:

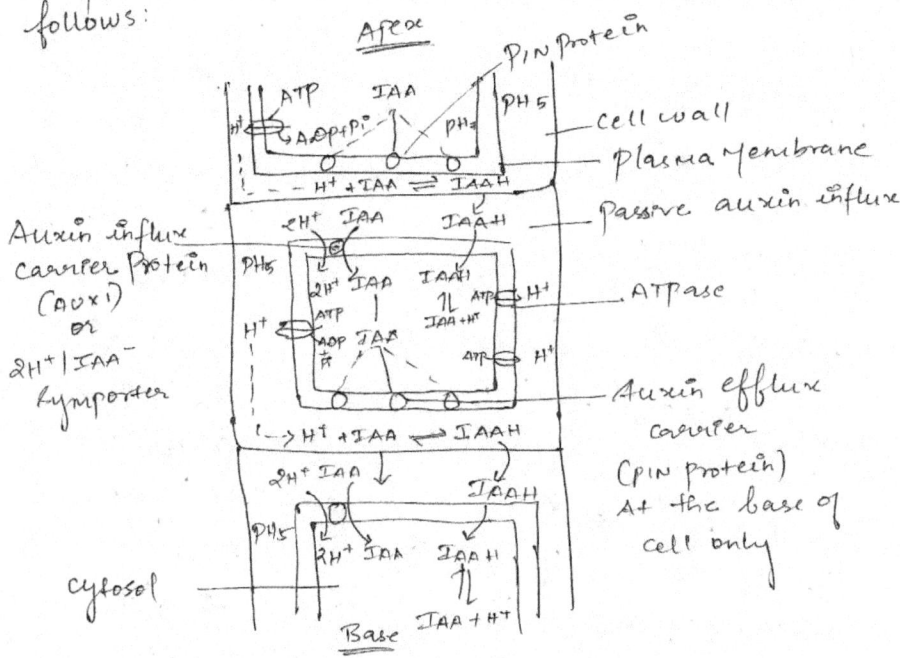

There are two main features of chemo-osmotic Model:
1) Auxin influx
2) Auxin efflux

Auxin influx
→ IAA may exist in 2 forms, one is protonated

on undissociated form (IAAH) which is highly lipophilic and can cross plasma membrane easily. Other form is dissociated or anionic form (IAA⁻) that does not cross plasma membrane unaided.

→ Under low pH (acidic), IAAH form predominates such as that exist in the cell wall space. Low apoplastic pH of about 5 is maintained due to the activities of ATPase which is present all around the plasma membrane. IAA⁻ which may be present in the cell wall combine with H⁺ to form IAAH that can passively diffuse across the plasma membrane.

→ IAA⁻ (anionic form) may also cross the plasma membrane from cell wall by secondary active co-transport mechanism via 2H⁺/IAA⁻ symporter (influx carrier protein) that is uniformly distributed around the cell. The influx carrier proteins are called as AUX1 and were first identified by Bennett et al in 1996 from the roots of Arabidopsis.

Auxin Efflux
→ The pH of cytosol is comparatively higher i.e. about 7 as a result of which IAAH

dissociates into H+ and IAA. In this dissociated form, auxin predominates the cytosol. In the dissociated form, the IAA⁻ exits the cell only through basally located auxin efflux carriers called as PIN proteins. The exit of IAA⁻ from the cell is driven by inside negative membrane potential.

If pH is high in the cytosol, the reversal of IAAH dissociation occurs. All these reactions are repeated in every cell for the transport of IAA through pH until the IAA reaches the cells in the site of action.

Destruction or inactivation of Auxin in plant

After the activation of growth, IAA is destroyed in the cells of target tissue. The chief method of degradation of auxin in plant is its oxidation by O_2 in the presence of the enzyme IAA-oxidase or peroxidase. Oxidation involves removal of CO_2 from the carboxylic group of auxin (IAA) and results in the formation of a variety of compounds, but 3-Methylene-oxindole is the major end product.

IAA + CH_2COOH + H_2O_2 + O_2 $\xrightarrow{\text{IAA oxidase (Peroxidase)}}$ CH_2 + H_2O + CO_2

Physiological roles of auxin

The various physiological roles performed by auxin are as follows:

→ Auxins induce elongation of plant cells and initiate cell division. It is responsible for callus formation in tissue culture.

The auxin causes cell elongation probably by any of the following mechanisms:
- By increasing osmotic solutes of the cells
- By reducing wall pressure
- By increasing the permeability of cells to water
- By an increase in cell wall synthesis
- By inducing the production of specific DNA dependent new m-RNA and specific enzymic proteins.

→ Auxin is responsible for apical dominance in which the apical bud is dominant in growth and does not allow the growth of lateral buds
→ Auxin initiates the formation of root
→ Auxin promotes cambial action
→ Auxin is used to break seed dormancy
→ Auxin initiates early flowering and fruiting
→ IAA induces Parthenocarpy
→ Auxin inhibit leaf fall and fruit fall.

Plant Growth and Growth regulators

→ Auxins are used as weed killers.
→ Auxin stimulates respiration.
→ Auxin induce the formation of Parthenocarpic fruits.
→ Auxin induces vascular differentiation in plants.

Gibberellins

Gibberellins are plant growth hormones which promote elongation of internodes. They occur naturally in all plants. The concentration of gibberellin is higher in reproductive cells when compared to vegetative cells.

History of Discovery

In the early 1912, Sawada found that some substances secreted by Gibberella fujikuroi in rice plant are responsible for the elongation of internodes.

Another young Japanese scientist Kurosawa had been working to find out the reason for the tall growth, thin and pale structure of rice seedlings infected by the fungus G. fujikuroi. These are the symptoms of 'Backanae disease'. In 1926, Kurosawa succeeded in obtaining a filtered extract of this substance (fungus).

In 1935, Yabuta isolated the active substance which was heat stable and gave it the name 'gibberellin'.

Yabuta and Sumiki in 1938 isolated gibberellin in crystalline form and identified gibberellin-A and gibberellin-B from their original Preparation. But the European and American scientists didn't gave importance to this because many scientists were working on growth hormones and English translation of Japanese work were not available until World War II.

In 1950 Mitchell at the Biological Welfare Centre in USA and Stodola in 1955 at the US Department of Agriculture were engaged to isolate this substance on commercial basis, thus the importance of gibberellins were made realized by the western world.

In 1955, Brian et al in England independently obtained pure sample of a single gibberellin at Imperial Chemical Laboratories and named it as gibberellic acid. The structure was established by cross et al in 1961.

At present, about 125 types of gibberellins have been isolated and chemically characterized. They are named as GA_1, GA_2, GA_3, GA_4 etc. Of these GA_1 and GA_{20} are widely distributed in higher plants. GA_1 bring out shoot elongation in higher plants. The most commercially available Gibberellin is GA_3 which is applied in many plant breeding programs.

Structure of Gibberellins

Gibberellins are tetracyclic diterpene acids having an ent-gibberellane ring skeleton in their structure. The four rings are often termed as A, B, C and D.

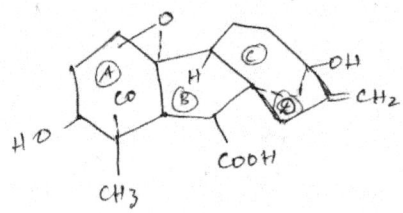

A ring is hexagonal in shape and is composed of 6 carbons. The ring B is pentagonal in shape and is composed of 5 rings. Ring C is hexagonal and Ring D is pentagonal. There is a COOH group at the 7th carbon atom of gibbane ring. Most of the gibberellic acid contain 20 carbon atoms in the gibbane ring but some of them only have 19 carbon atoms. Gibberellins having 19 carbon atoms are more biologically reactive than those containing 19 carbon atoms.

Biosynthesis of Gibberellins in plants

Gibberellins are structurally related to terpenoids. The terpenoids are synthesized from

Isopentenyl Pyrophosphate (IPP) which in turn is synthesized from acetate. Hence the primary precursor for gibberellin synthesis is acetate.

Fig.- Biosynthesis of gibberellins from acetate

$$\text{Acetate}$$
$$\downarrow \leftarrow CoA$$
$$\text{Acetyl-CoA}$$
$$\downarrow \substack{\leftarrow \text{Acetyl CoA} \\ \rightarrow CoA}$$
$$\text{Aceto-acetyl-CoA}$$
$$\downarrow \substack{\leftarrow \text{Acetyl CoA} \\ \rightarrow CoA}$$
$$\text{Hydroxy glutaryl CoA}$$
$$\downarrow \substack{\leftarrow 2NADPH_2 \\ \rightarrow 2NADP}$$
$$\text{Mevalonic acid}$$
$$\downarrow \substack{\leftarrow 2ATP \\ \rightarrow 2ADP}$$
$$\text{Mevalonic acid Pyrophosphate}$$
$$\downarrow \substack{\leftarrow ATP \\ \rightarrow ADP+Pi}$$
$$CO_2 + \text{Isopentenyl Pyrophosphate (IPP)}$$
$$\downarrow \text{Isomerization}$$
$$\text{Dimethyl allyl Pyrophosphate}$$
$$+IPP \downarrow \substack{\text{condensation} \\ \rightarrow PP}$$

Geranyl Pyrophosphate
+IPP ↓ condensation
↳ PP

Farnesyl Pyrophosphate
+IPP ↓ condensation
↳ PP

Geranyl geranyl Pyrophosphate
↓ inhibited by Amo-1618, CCC & phosfon D

Copalyl-Pyrophosphate
↓ inhibited by phosfon-D

ent-Kaurene
↓ inhibited by Paclobutrazol

GA_{12} Aldehyde
↓
GA_{12}
↓
All other GAs ← GA_{53}

In plants GAs are biosynthesized in apical tissues and there are 3 main sites of their biosynthesis.
→ Developing seeds and fruits
→ Young leaves of developing apical buds
→ elongating shoots

The pathway of GA biosynthesis can be divided into 3 stages each of which is accomplished in a different cellular compartment.

Stage-1 Synthesis of Terpenoid Precursors

The building block of gibberellin, the isopentenyl diphosphate (IPP) is synthesized from glyceraldehyde 3 phosphate and Pyruvate. In endosperm of seeds, IPP is produced in cytosol from Mevalonic acid which is derived from acetyl CoA.

IPP (5C) units are added together one by one to produce intermediates geranyl diphosphate (10C), farnesyl diphosphate (15C) and geranyl geranyl diphosphate (20C). The geranyl geranyl diphosphate (GGPP) is converted to ent-Kaurene by a cyclic reaction in the proplastids of Meristems.

The synthesis of these precursors is inhibited by AMO-1618, cycocel and Phosphon D.

Stage-2 Oxidation of ent-Kaurene

Methyl group on ent-Kaurene is oxidized to carboxylic acid. B-ring is then contracted from 6 to 5 carbon ring to form GA_{12}-aldehyde.

GA12 aldehyde is then oxidized into the first gibberellin GA12 in the presence of cytochrome P450 and monooxygenases. This takes place in endoplasmic reticulum. GA12 is the precursor to all other GAs. Hydroxylation of C13 of GA12 results in the formation of other gibberellins (GA53).

Stage-3 Synthesis of other Gibberellins

Synthesis of all other gibberellins takes place by a group of dioxygenases in the cytoplasm. These enzymes use 2-oxoglutarate and O_2 as co-substrates and Fe^{2+} and ascorbate as co-factors. These enzymes carry out hydroxylation at carbon 13 of GA12.

Distribution of Gibberellins in plants

Gibberellins have been found from both phloem and xylem exudates from a variety of plants. The gibberellins are found in all parts of higher plants including shoots, roots, leaves, flowers, petals, anthers and seeds. Gibberellins are also reported in plastids. The reproductive parts contain much higher concentration of gibberellins than the vegetative parts.

Transport of gibberellins in plants

The transport of gibberellins in plants is non-polar. Based on the flow pattern, it is believed that gibberellins are translocated through phloem which is similar to that of carbohydrates and organic solutes. The transport of gibberellin may also occur in xylem due to its lateral movement between the xylem and phloem. The gibberellins are translocated in plants in their bound form as gibberellin glycosides.

Deactivation of Gibberellins

Several mechanisms are available for the deactivation of gibberellins.

→ Introduction of 2-β-hydroxyl group into GA markedly reduces the biological activity.

→ Conversion of free GAs into their bound forms like gibberellin-glycosides or gibberellin-glycosyl ethers or esters.

Physiological Roles of Gibberellins

→ Gibberellins promote the elongation of internodes and thereby promoting the elongation of stem. Gibberellins induce extension of cell wall without acidification thus producing xyloglucan transglycosylase (XGT). XGT promote the loosening of cell wall leading to stretching. Gibberellins too induce the expression of various genes like CDC2, CDKs and cyclins genes for the production of proteins needed for cell division and suppress genes like GAI, RGA and SPY which interfere the cell wall promoting Proteins.

→ Gibberellins promote the growth of long day plants under short day conditions. This is termed as 'bolting'.

Some plants (Rudbeckia speciosa, Hyoscyamus niger) grow well and produce flowers only under long day conditions. In short day conditions, these plants assume a rosette form and never produce flowers. The exogenous application of gibberellin the dwarf plants grow as a tall plant and produce flowers under short day conditions.

→ Gibberellins have little direct effect on the growth of root. Gibberellin treatment to shoot of dwarf plants enhances both shoot and root growth.

→ Many woody plants come to flowering stage

only after a period of juvenile growth. The gibberellin treatment extends the juvenile period and delays the flowering.

→ Gibberellin treatment to some juvenile conifers induced early flowering by reducing the juvenile phase of growth

→ GA_4 and GA_7 are effective in inducing early flowering.

→ Exogenous application of gibberellins induces fruit setting in some plants.

→ Seedless and fleshy tomatoes and grapes are also produced by gibberellin treatments.

→ Seed germination involves the activation of embryo growth and weaking of growth-constraining endosperm layer and the mobilization of food reserves of endosperm to seedling.

Gibberellin stimulates the production of hydrolases enzymes to solubilize food reserves in the endosperm and to mobilize them to growing seedlings. Thus, gibberellins promote germination. Steps involved are;

→ Gibberellin synthesized in coleoptile and scutellum and transported to starchy endosperm.

➡ In the endosperm, gibberellins bind with GA receptor on the surface of the cells of aleurone layer

➡ Inside cytosol, GA binds with GA-signalling intermediate and enters the nucleus.

➡ In the nucleus, it binds to DELLA and SPY genes to repress their expression. GA-MYB and some other genes are induced to express their proteins

➡ GA-MYB proteins induce the expression of mRNA of α-amylase

➡ This mRNA comes out of the nucleus and enters the ER where it is translated into α-amylase.

➡ This α amylase is released in the form of coated vesicle from golgi apparatus

➡ With the help of Ca^{++} dependent calmodulin and protein kinase activity, the vesicles containing α-amylases are released into the endosperm cells.

➡ In the endosperm cells, α-amylase hydrolyzes starch into sugars which are then transported to the growing tissues of embryo.

→ In temperate plants, buds produced in the autumn remain dormant, because of severe cold, until the spring season comes. In these cases, bud dormancy can be broken by the exogenous application of gibberellins.

Cytokinins

Cytokinins are plant growth hormones whose physiological effect is to promote cell divisions. Chemically cytokinins are kinins which are purine derivatives. They are considered as the degraded products of adenine.

History of discovery

Growth promoting effect of cytokinins was first discovered by Miller et al in 1955 who were working in Prof. Skoog's lab at the University of Wisconsin on the growth of tobacco pith callus in culture and wanted it to grow indefinitely. They added many growth substances, nutrients, vitamins etc. into the culture medium but were failures. They noticed an old bottle of DNA kept for several years in the laboratory. They observed that the tobacco pith callus could grow for longer period on adding the content of the bottle to the culture medium. Similar results were obtained with yeast extract.

They isolated this substance by autoclaving the DNA which had been stored for long. It could be easily precipitated by silver salts and was soluble in 90% alcohol, indicating that possibly it was a purine compound. Later on, it was identified as 6-furfurylamino purine.

The term cytokinin was proposed by Letham in 1963. Cytokinins are also termed as phytokinins. Roots, fruits, endosperm and embryos contain large amount of cytokinin.

Chemical Structure

Kinetin

Chemical studies of Miller et al in 1955 confirmed that the molecule is an adenine derivative, 6-furfurylaminopurine.

Fig: Structure of Kinetin Molecule

Kinetin is not a naturally occurring plant

growth regulator and is formed during the heat induced degradation of DNA. The deoxyribose sugar of adenosine is converted to a furfuryl ring and then it is shifted from 9th position to 6th position of the adenine.

Zeatin

Zeatin is the most common cytokinin present in higher plants and some bacteria. Letham in 1963 discovered that extracts of immature endosperm of corn (Zea mays) were having same effect as Kinetin. In 1973, Letham isolated the substance and identified it as trans-6-(4-hydroxy-3-Methyl-2-butenyl amino) Purine. It is now popularly termed as Zeatin.

Fig: Structure of zeatin

The structure of zeatin is similar to that of Kinetin. Both are adenine molecules or amino-Purine derivatives. In both the cases, the side

Plant Growth and Growth regulators

chain is attached to 6th nitrogen of the aminopurine. But the side chain of Zeatin has a double bond and hence it can exist in the cis or trans configuration.

The cis and trans configurations of zeatin are inter convertible by the enzyme Zeatin isomerase. The trans form is more biologically active than cis form.

Synthetic and Natural cytokinins

→ Kinetin, Zeatin, N_6-(Δ^2-Iso-pentenyl)-adenine (ip) and Dihydrozeatin (DZ) are naturally occurring cytokinins in plants.

→ Synthetic compounds also function like Zeatin and are termed synthetic cytokinins. They include Benzyladenine (benzylaminopurine), N, N_2-Diphenylurea and Thediazuron.

Biosynthesis of Zeatin and other Natural cytokinin

Cytokinins are synthesized from adenosine Monophosphate (AMP) and isopentenylpyrophosphate by condensation reaction in the presence of enzyme isopentenyl transferase. The product of this condensation reaction is N^6-(Δ^2-Isopentenyl)-

adenosine-5'-Monophosphate which is supposed to be the precursor to all other natural cytokinins.

Fig: Biosynthesis of cytokinins

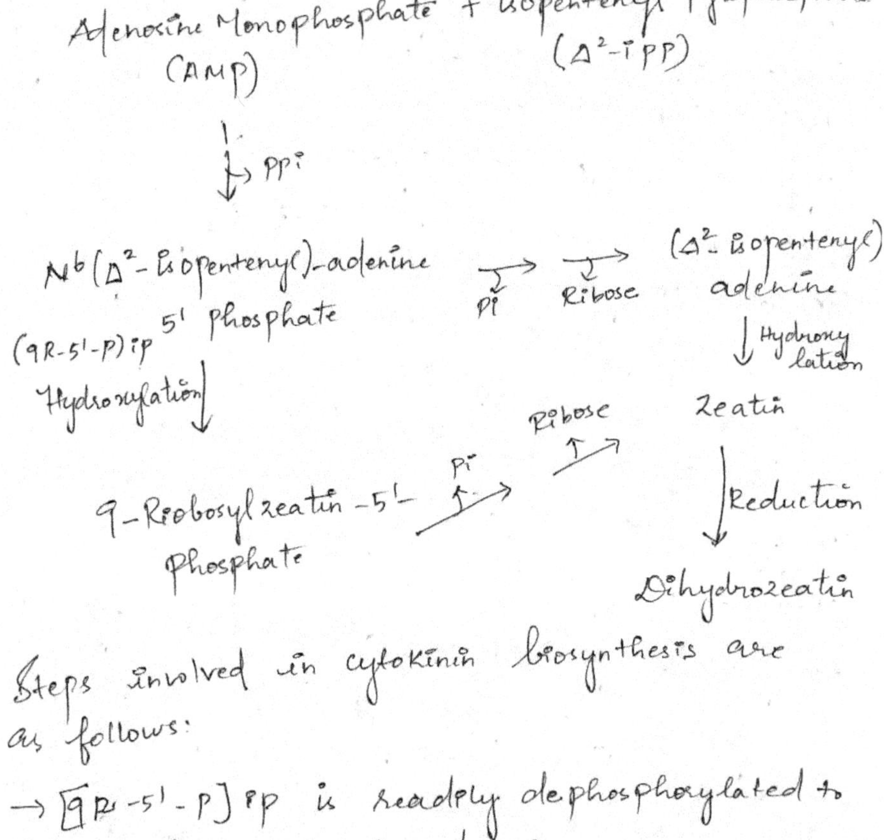

Adenosine Monophosphate (AMP) + Isopentenyl Pyrophosphate (Δ^2-iPP)

↓ PPi

N^6(Δ^2-Isopentenyl)-adenine 5' phosphate (9R-5'-P)iP
↓ Hydroxylation

9-Ribosyl zeatin-5'-Phosphate

→ Pi → Ribose → (Δ^2-Isopentenyl) adenine
↓ Hydroxylation

Ribose → Zeatin
↓ Reduction
Dihydrozeatin

Steps involved in cytokinin biosynthesis are as follows:

→ [9R-5'-P]iP is readily dephosphorylated to form N^6-(Δ^2-Isopentenyl)-adenosine.

→ From N^6-(Δ^2-Isopentenyl)-adenosine, ribose

Sugar is removed, so that N^6-(Δ^2-isopentenyl)-adenine is formed

→ Isopentenyl side chain of ip is now hydroxylated to form free zeatin

→ [9R-5'-P] ip may be hydroxylated directly to give 9-ribosyl-zeatin-5'-phosphate [ie (9R-5'p)z]. The phosphate group and then the ribose sugar are removed from [9R-5'P]z in sequence to form free zeatin

→ Reduction of double bond in isopentenyl side chain of zeatin would yield dihydrozeatin (diHZ)

Distribution and Transport of cytokinins in Plants

Cytokinins are not widely distributed in different plant parts. Cytokinins are produced in roots mainly during seedling stage and translocated to shoots and leaves of the plant. Highest concentrations of cytokinins are observed in root and shoot tips. Cytokinins are found in xylem exudate and appear to be translocated through xylem. The cytokinins are transported in their bound form as ribosides in plant. It is found that a signal from shoot regulates the transport of zeatin ribosides from roots.

Deactivation of Cytokinins

There are 2 main ways for inactivating cytokinins in plants 1) by conjugation and 2) by oxidation.

→ **By conjugation (Reversible or irreversible)**

The regulation is carried out by conjugation of cytokinin with either glucose or amino acids which may be reversible or irreversible depending upon the nature of conjugation.

→ **By oxidation (Irreversible)**

In many plant tissues, cytokinins are irreversibly inactivated by the enzyme cytokinin oxidase which cleaves the isopentenyl side chain of cytokinins.

iP →[cytokinin oxidase, O_2]→ Adenine + 3-Methyl 2-butenal

Physiological role of cytokinins

→ Cytokinins are found to induce cell division in both roots and shoots. Therefore cytokinins along with auxins play an important role in the division of Meristematic cells in adult plants.

→ For proliferation of shoot meristem the level of cytokinin must be high whereas for the proliferation of root meristem, the level of cytokinin must be low.

→ Cytokinins take part in the regulation of cell division cycle. Both auxin and cytokinins participate in the regulation of the cell cycle by controlling the activity of cyclin dependent Kinases (CDKs)

→ CDKs in CDCs in higher levels permit the cells to divide in the presence of adequate amount of auxins.

→ Cytokinins induce the enlargement of cells in leaves. It is also found that Kinetin induces the formation of interfascicular cambium (in pea stem)

→ Cytokinins are very important in inducing the growth of lateral buds in plants when

the apical bud remains intact with the plant
- Cytokinins bring about delay of leaf senescence
- Cytokinins induce the movement of nutrients into the leaves from other parts of the plant. This phenomenon is termed as cytokinin-induced nutrient mobilization.
- In the dark-grown seedlings, the hypocotyl and internodes are more elongated, cotyledons and leaves do not expand and chloroplast do not mature. Such plants are called etiolated plants. If the etiolated leaves are sprayed with cytokinin before exposing them to light, the etioplasts develop more extensive grana, chlorophylls and photosynthetic systems.
- Seeds of some light sensitive plants do not germinate unless they are stored in darkness. Such seeds can be made to germinate without dark treatment by soaking the seeds in kinetin solution for sometime.

Ethylene

Ethylene is also termed as gaseous hormone. It is a ripening hormone which is produced in trace amounts in almost all plant cells. A unique feature about ethylene is that it is an exogenous chemical affecting plant growth.

History of Discovery:

Ethylene as a plant product was discovered by Cousins in 1910. He made the first indication that the ethylene might be a natural product of plant tissues. He mentioned that "bananas should not be stored with oranges in ships because some emanations from oranges caused banana to ripen prematurely.

It was R. Gane in 1934 who clearly stated that ethylene is actually a natural product of ripening fruits and is responsible for faster ripening process.

However, the discovery of ethylene as a plant growth regulator may indirectly be traced back to 19th century when street lights were used to be lighted with illuminating coal gas.

It was observed that the trees in the vicinity of the street lamps defoliated more extensively than other trees. Even ancient Chinese knew that their harvested fruits would ripen much faster in burning incense. Later, it was identified that ethylene is an important component of coal gas and burning incense which was probably the reason for such effects on plants.

The credit for first establishing the fact that ethylene affects plant growth goes to a Russian physiologist Dimitry N. Neljubow who in 1901 identified ethylene in laboratory air from illuminating coal gas which caused typical symptoms in etiolated pea seedlings grown in the dark in the laboratory.

⇒ Inhibition of stem elongation
⇒ Stimulation of radial swelling of stems
⇒ Horizontal growth of stems with respect to gravity.

These symptoms were later termed as 'triple response'.

Chemistry

Ethylene (C_2H_4) is the simplest olefin with molecular weight 28.

$$\begin{array}{c} H \\ \diagdown \\ C=C \\ \diagup \diagdown \\ H H \end{array} \quad \text{(or)} \quad CH_2 = CH_2$$

Ethylene is flammable and highly volatile that readily undergoes oxidation to form ethylene oxide. The ethylene can be fully oxidised to CO_2 through ethylene oxide in many plant tissues. Ethylene is readily absorbed by Potassium permanganate ($KMnO_4$). This is used frequently to remove excess ethylene from the storage chambers.

Biosynthesis of Ethylene

Ethylene is found to be synthesized in plant tissues from the amino acid Methionine. A non protein amino acid, 1-amino cyclopropane-1-carboxylic acid (ACC) is an important intermedia and also immediate precursor of ethylene biosynthesis. The two carbons of ethylene molecule are derived from carbon 3 and 4 of Methionine.

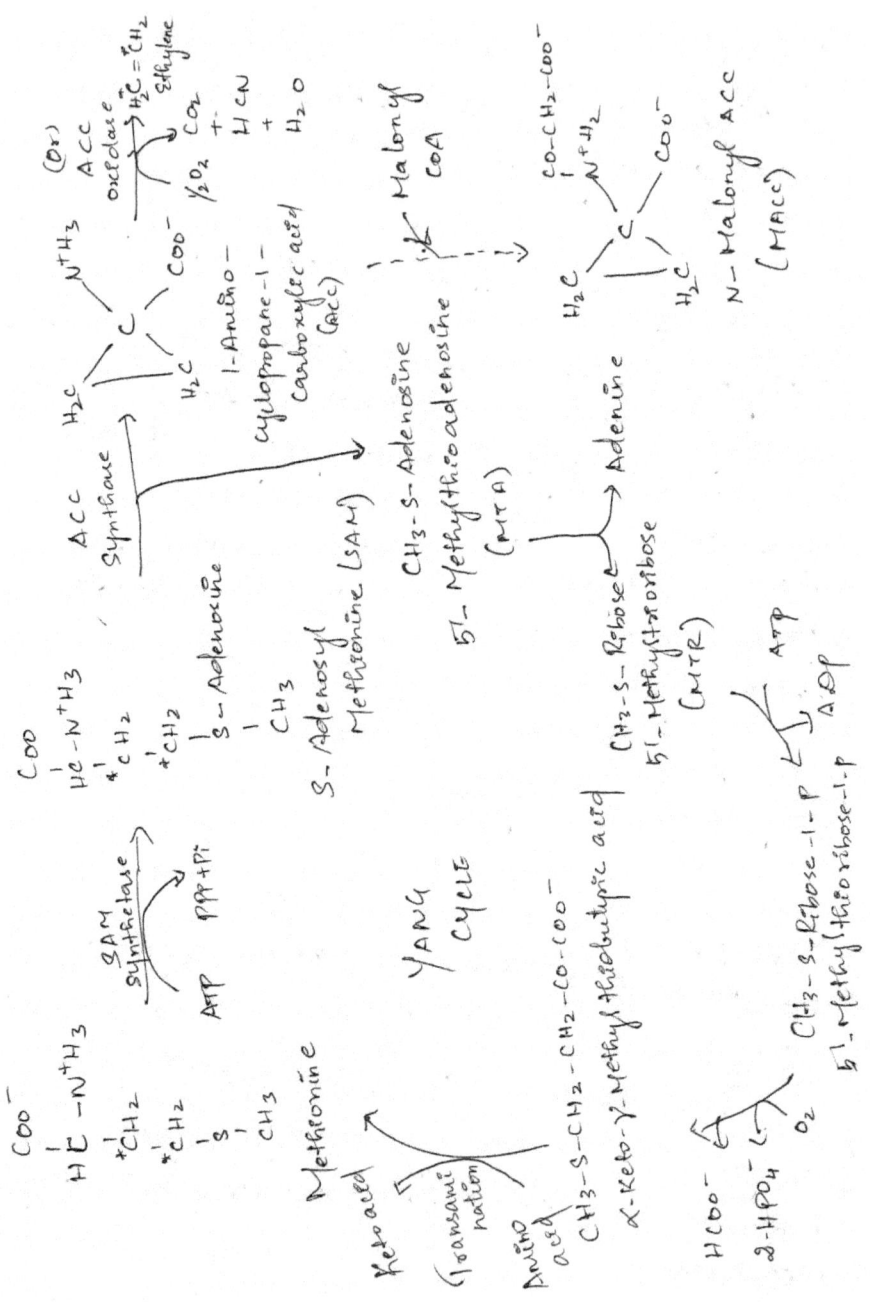

Steps involved in biosynthesis are as follows:

Step-1: An adenosine group is transferred to Methionine by ATP to form S-adenosyl Methionine (SAM) catalysed by the enzyme SAM-Synthetase (i.e. Methionine adenosyl transferase)

Step-2: SAM is cleaved to form 1-amino cyclo propane-1-carboxylic acid (ACC) and 5'-Methyl thioadenosine (MTA) by enzyme ACC-Synthase. The rate limiting step in ethylene biosynthesis is the production of ACC. Exogenously supplied ACC greatly enhanced the production of ethylene in plant tissues.

Step-3: ACC is oxidised by the enzyme ACC-Oxidase to form ethylene. Two molecules HCN and H_2O are eliminated.

Yang cycle

There is only limited amount of free Methionine in plant tissues. Hence, to sustain the normal rate of ethylene biosynthesis, the sulphur released during ethylene biosynthesis is recycled back to Methionine again via Methionine cycle or Yang cycle.

Distribution of Ethylene in Plants

Ethylene is produced by all groups of plants including bacteria, fungi, BGA, bryophytes, Pteridophytes, gymnosperms and angiosperms. Ethylene is produced in all plant parts in higher plants like roots, stems, leaves, tubers, bulbs, fruits and seeds. But, the production varies based on the type of tissue and stage of development. It is highest in senescing tissues and ripening fruits. Ethylene is biologically active at very low concentration (<1ppm).

As a by-product of hydrocarbon combustion, ethylene is also a common environmental pollutant that can play havoc with green house cultures and/or laboratory experiments.

Transport of Ethylene in Plants

Because of its hydrophobic nature, ethylene can easily pass through plasma membrane into the cell, easily diffuse within the plant or plant part are flushed out of plant tissues through intercellular spaces. Cuticle acts as resistant barrier, whereas stomata, lenticels and cut spaces serve as exit points for ethylene

Deactivation of ethylene

Due to its high diffusivity, excess ethylene can be readily flushed out of plant tissues. CO_2, ethylene oxide and ethylene glycol are found to be major breakdown products of ethylene in plant tissues.

Physiological role of ethylene

→ Ethylene promotes the ripening of fleshy fruits hence it is popularly called as ripening hormone. The fruits which are not responsive to ethylene action are termed as climacteric fruits.

→ Ethylene and high amount of auxin are needed to bring out epinasty. Epinasty is the phenomenon of downward curvature of leaves due to the fast growth of the upper side of petiole.

→ Ethylene causes 'triple response' of etiolated seedlings (pea) which consists of inhibition of stem elongation, stimulation of radial swelling of stems and horizontal growth of stems with response to gravity (i.e. diageotropism)

→ Ethylene promotes lateral cell expansion

→ Ethylene induces the formation of plumular

hooks. Hook like bending of shoot tip of etiolated dicot seedlings is called plumular hook.

→ Ethylene is known to inhibit linear growth of roots of dicotyledonous plants

→ Dormancy of seeds and buds can be broken by ethylene treatment. Ethylene initiates seed germination in cereals, strawberry and apple.

→ Ethylene promotes stem and petiole elongation in submerged aquatic plants (Eg: *Ranunculus sceleratus*)

→ Ethylene is known to induce adventitious root formation from leaves, stems, flower stalks and roots. It is also a positive regulator of root hair formation in many plants.

→ Ethylene inhibits flowering in many plants, but it promotes flowering in pineapple and its relatives.

→ Ethylene or ACC treatment accelerates leaf senescence

→ Ethylene increases the abscession of leaves, flowers and fruits.

Abscisic acid (ABA)

Abscisic acid is a growth hormone which affects the plant growth.

History of Discovery

Carns and Addicott in 1963 discovered it from young cotton fruits. They isolated a substance strongly antagonistic to growth and named it as Abscesin II. Later on the name was changed to Abscisic acid.

Eagles and Wareing in 1963 & 1964 pointed out the presence of a substance in Betula pubescens leaves which inhibited growth and induced dormancy of buds and therefore named it as 'dormin'. Later by the work of cornforth etal in 1965, it was found to be identical with abscisic acid.

Chemistry

Chemically, ABA is a 3-Methyl 5-1' (1'- hydroxy,

4-oxy 2', 6', 6'-trimethyl-2-cyclohexane-1-yl). Cis, trans-2, 4-penta-dienoic acid. It is a 15-carbon sesquiterpene. It has a cyclohexan ring, one keto group, one hydroxyl group and a terminal carboxylic acid side chain in its structure. ABA resembles terminal portion of some carotenoids like violaxanthin and neoxanthin. It also appears to be the breakdown product of such carotenoids.

Any change to its molecular structure results in the loss of its activity. It occurs in cis, trans isomeric forms that are decided by the orientation of −COOH group around 2nd carbon atom in the molecule.

The cis form is biologically active in plants. The trans form is inactive but can be interconvertible with cis-ABA.

Biosynthesis of ABA in plants

ABA is synthesized in higher plants indirectly through carotenoid pathway as breakdown product of 40-c xanthophyll, such as violaxanthin or neoxanthin.

Fig:- Steps Involved in ABA synthesis

Isopentenyl diphosphate (IPP)
(5C)
↓
Farnesyl diphosphate (FPP)
(15C)
↓
Zeaxanthin (40C)
↓ Zeaxanthin epoxidase
Trans - Violaxanthin (40C)
↓
9'- cis- Neoxanthin (40C)
O_2 ↓ NCED
Xanthoxal (15C)
(xanthoxin)
↓
ABA- aldehyde (15C)
↓
Abscisic acid (15C)
(ABA)

Steps involved in ABA biosynthesis are as follows:

→ Initial steps of ABA biosynthesis take place in Chloroplasts or other plastids while final steps occur in cytosol

→ Violaxanthin is produced from zeaxanthin catalysed by enzyme zeaxanthin epoxidase (ZEP).

→ Violaxanthin is converted into 9'-cis-neoxanthin which is cleaved into 15-C compound xanthoxal and a 25-C epoxy aldehyde in the presence of enzyme 9'-cis-epoxycarotenoid deoxygenase (NCED).

→ Xanthoxal is finally converted to ABA in cytosol in 2 oxidation steps in the presence of aldehyde oxidases. Abscisyl aldehyde is formed as intermediate. The enzyme aldehyde oxidases need Mo as cofactor.

Distribution of ABA in Plants

ABA is an ubiquitous plant hormone in vascular plants. They are absent in liverworts and algae where in a compound named lunularic acid is present which appears to be possible functional equivalent of ABA.

Within the plant, ABA has been detected in all major organs or living tissues from root caps to apical buds. It is also noted in xylem, phloem and nectar. It is produced in all types of cells containing chloroplasts or other plastids. It is predominantly present in mature green leaves.

Transport of ABA in plants

ABA occur mostly in free form but it may also occur in conjugated form as glycoside with some simple sugar molecule such as glucose. ABA is biologically inactive in bounded or conjugated form.

Externally applied ABA enter the tissues rapidly and gets distributed freely in all directions within the plant. Cell to cell transport of ABA is slow and non-polar. ABA in xylem and phloem sap is most probably translocated throughout the plant via vascular tissue. ABA produced in root cap moves basipetally into the central vascular tissue.

Deactivation of ABA in plants

2 ways

→ Oxidation of phaseic acid → This is the main route of ABA degradation in plants. ABA is oxidised to phaseic acid with a subsequent reduction of keto group on the cyclohexane ring to form dihydrophaseic acid (DPA). In some, DPA is further metabolized to form 4' glucoside of DPA

→ **Conjugation as Glucosides**: Free ABA can be inactivated in plants by covalent conjugation with some simple sugar molecule like glucose to form ABA-β-D-glucosyl ester which gets accumulated in vacuole.

Physiological effects of Abscisic acid

→ Stomatal closure is enabled by ABA under water stress conditions. Hence it is called as stress hormone
→ Induce bud dormancy in some temperate zone trees
→ Induces dormancy of seeds require stratification
→ Induce process of tuberization
→ Induce senescence of leaves
→ Promote ripening of fruits
→ Bring out abscission of leaves, flowers and fruits
→ Increase the resistance of temperate zone plants to frost injury.

→ Inhibit the GA-induced synthesis of α-amylase in aleurone layers of germinating barley.
→ Inhibit precocious germination and vivipary.

Brassino Steroids

Brassinosteroids or brassins are group of steroids which have distinct growth promoting activity in some plants especially in the stems. These were first isolated in 1979 from bee collected pollen grains of *Brassica napus* (hence the name) chemically, the structure of brassinosteroid resemble to that of steroid hormones of animals.

Fig: Brassinolide

Physiological effects of Brassinosteroids resemble to that of auxins, except that brassinosteroids could affect cell elongation and gene expression quite independently of auxins. They also play significant roles in many light and hormone regulated Processes like expression of light regulated genes, the Promotion of cell elongation, leaf and chloroplast senescence and floral induction.

Morphactins

Morphactins are a new group of synthetic growth regulators came into action in 1960s. The term 'Morphactins' refers to 'Morphologically active substances'. Their action is mainly inhibitory to development and growth.

Chemically, Morphactins are synthetic derivative of fluorene-9-carboxylic acid. Absorbed through seeds, leaves or roots and gets distributed in the plants basipetally and acropetally.

Physiological effects

→ General inhibition of internodes elongation
→ Reduction in laminar area
→ Reduction of apical dominance
→ Inhibit lateral root formations
→ Abolition of phototropism of shoots and geotropism of roots.

Traumatic acid (Wound hormone)

Haberlandt in 1913 postulated that the injured cells release a 'wound hormone' which causes the adjacent uninjured cells to become Meristematic and divide until the wound is healed up.

Englsch et al in 1939 isolated and identified this substance as Traumatic acid

Chemistry

$$COOH-CH=CH-(CH_2)_8-COOH$$

It is a straight chain unsaturated dicarboxylic acid. Traumatic acid requires glutamic acid, sucrose and phosphates as cofactors for its activity to form callus.

References

- Plant Physiology – Robert M. Devlin
- Plant Physiology – Dr. V. K. Jain
- Plant Physiology – Vladimir Palladin
- Plant Physiology – Dr. Annie Ragland et al.
- Plant Physiology – Subhas Chandra Datta
- Plant Physiology – Nikolai Aleksandrovich
- Plant Physiology – Taiz & Zeiger

www.ingramcontent.com/pod-product-compliance
Lightning Source LLC
Chambersburg PA
CBHW071422220526
45469CB00004B/1388